Lichen Study Guide

for Oklahoma and Surrounding States

Lichen Study Guide

for Oklahoma and Surrounding States

Sheila A. Strawn

BRIT
PRESS

Lichen Study Guide for Oklahoma and Surrounding States

Printed in Korea
ISBN: 13-978-1889878-55-3

Please cite this book as follows:
Strawn, S.A. 2017. *Lichen Study Guide for Oklahoma and Surrounding States.* Sida, Bot. Misc. 48. Botanical Research Institute of Texas, Fort Worth, Texas, U.S.A.

Production: Rebecca Horn, rhorngraphics, rlrchorn@verizon.net
Cover Design and photos: Sarah Hearn
Inside Photography by Sheila Strawn
Additional photographs by Robert Harris and Caleb Morse
Back Cover and Frontispieces: *Cladonia cristatella*, *Ramalina* sp., *Lichinella cribellifera*, *Teloschistes chrysophthalmus* & *Parmotrema* sp., *Teloschistes exilis*, *Lichinella nigritella*, *Graphis scripta*, and *Cladonia subtenuis.* Photos by Sarah Hearn.

Distribution of copies by:
Botanical Research Institute of Texas Press, 1700 University Dr., Fort Worth, Texas 76107-3400, USA
Telephone: 1.817.332.4441; Fax: 1.817.332.4112; www.britpress.org; press@brit.org; shop.brit.org

BRIT
PRESS
2017
Sida, Botanical Miscellany, No. 48
ISSN: 0833-1475
15 May 2017

Contents

Acknowledgments

Steven Strawn, my field and lab partner, even before we began studying lichens. We have attended workshops together and lead workshops together. He is always ready to identify a box of lichens or pack up the gear and go collect specimens.

Sarah Hearn, artist, photographer, and lichenologist. She donated the specimens from her public art project, "Symbiotic Cooperation" to the CSU Herbarium at the University of Central Oklahoma where the Strawn Lichen Collection is housed. She photographed many lichens in the collection, provided comments to improve the book, and composed the front cover for this guide.

Robert Harris, Texas lichen enthusiast who made valuable comments for this guide as he helped field test it at several workshops. He also provided several photographs illustrating lichen thallus components.

James Lendemer, Assistant Curator for the Institute of Systematic Botany at New York Botanical Garden. He has provided valuable information and insight for identifying lichens. He identified many specimens that we had given up on, gave patient advice at Tuckerman Workshops we attended, and gave very helpful reviews of the manuscript.

Irwin Brodo, author of *Lichens of North America*. He gave valuable advice and comments on two drafts of the manuscript. He taught the "Crustose Lichens" course for the Eagle Hill Field Seminars which Steve and I took at Humboldt Field Research Institute in Steuben, Maine in 2006.

Robert Egan, Professor Emeritus, University of Nebraska at Omaha (UNO). He performed TLC tests on many of our lichens and identified many more species. He gave us permission to access his UNO lichenology course website where I learned much of what I know about lichens.

Caleb Morse, Collections Manager of the Ronald L. McGregor (KANU) Herbarium at the University of Kansas in Lawrence. He patiently identified many, many specimens for us and gave advice about how to identify others. He suggested lichen resources that we use and recommend to other lichenologists. He also provided the excellent photograph (p. 43) of lichen conidia for this guide.

Richard Harris, Curator of Lichens at the New York Botanical Garden (NYBG). He identified numerous specimens for us and gave valuable field advice during Tuckerman Workshops. With support from the National Science Foundation, he and Doug Ladd authored *Ozark Lichens, Enumerating the Lichens of the Ozark Highlands of Arkansas, Illinois, Kansas, Missouri, and Oklahoma.* He provided us with a preliminary draft, which we have used extensively to identify lichens from eastern Oklahoma. He also gave a very helpful review of the manuscript.

Thomas Nash, co-author of *Lichen Flora of the Greater Sonoran Desert Region.* He reviewed many of our early species identifications from photos we brought to Tuckerman Workshops. He suggested tests or additional characters we could use to determine many specimens. His three-volume lichen flora of the arid southwest region was invaluable for identifying lichens in western Oklahoma.

William R. Buck, Senior Curator for the Institute of Systematic Botany at New York Botanical Garden. He reviewed the final version of the manuscript and provided critically important corrections.

Douglas Ladd, from the Nature Conservancy Missouri Field Office and co-author of *Ozark Lichens.* He offered advice and sources for information as we began learning about lichens in 2002.

Patricia Folley, volunteer taxonomist in the Bebb Herbarium (OKL) at the University of Oklahoma and personal friend. It was her suggestion and encouragement that convinced me to begin a study of *Lichens of Oklahoma.*

Lichen Study Guide

for Oklahoma and Surrounding States

Introduction

Oklahoma is a mixed region of many diverse habitats. It is known to have more plant species than other states its size. It would not be surprising to find the same is true about lichens, but there has yet to be a thorough statewide survey. Lichen biology is usually included in basic science textbooks as an example of a symbiotic association between particular fungi and cyanobacteria or green algae. There are, however, few opportunities to study the intricate structures and processes of these complex organisms compared to opportunities to study birds, rocks, flowers, and trees. Lichens are both intricate and fascinating. Their study is highly rewarding and you do not have to travel very far to find them. Once you know what to look for and where to look, you realize they are all around you. The purpose of this guide is to encourage large scale studies of lichens in this region by explaining the basic characteristics needed to understand lichen biology and identification.

Steve Strawn in the Wichita Mountains Wildlife Refuge northwest of Lawton.

About Lichen Study Guide for Oklahoma and Surrounding States

It is hoped that this guide will encourage the study of lichens in Oklahoma and in surrounding states with similar habitats. The lava-capped, higher elevation of Black Mesa State Park in the Oklahoma panhandle hosts lichen species also found in southeastern Colorado and eastern New Mexico. The habitats and substrates in the heavily wooded Ozark Mountains northeast of Tulsa and in southeastern Oklahoma's Ouachita National Forest and Lake Wister State Park will support species similar to those in eastern Kansas, southwestern Missouri, and western Arkansas. Woodland species in riparian and mixed grass communities along the Red River support many of the same foliose and fruticose lichens as those found in North Texas, while shortgrass communities at Black Kettle National Grassland and Washita Wildlife Refuge in western Oklahoma support many of the same soil lichens as those in the West Texas panhandle. The Tallgrass Prairie Preserve west of Bartlesville would have species similar to those in the tallgrass regions of Kansas and Nebraska. The uniqueness of Oklahoma lichens is found in places like the gypsum outcrops at Selman Living Lab south of Freedom, the mineral springs at Chickasaw National Recreation Area at Sulphur, and the granite rock of Quartz Mountains north of Altus and Wichita Mountains Wildlife Refuge west of Lawton. We need to study all these regions more extensively, and that is the most important reason for developing this study guide for lichens in Oklahoma and surrounding states.

We need to study all these regions more extensively, and that is the most important reason for developing this study guide for lichens in Oklahoma and surrounding states.

To build on this basic guide, additional resources are included in the Appendices. Suggestions for laboratory practice and literature searches are included in Appendix A. Even though this guide is based on lichens found in Oklahoma and surrounding states, the Practice Keys in Appendix B are intended to give the reader experience in understanding the basic characteristics used to identify lichens all over the world. The keys are based on common thallus features and chemical reactions, so that in a laboratory setting an instructor who is familiar with lichens in any region of the world would be able to substitute local species. Appendix C lists species used in each of the activities and keys.

When you are ready to identify lichens from the field, the appendices also include several more resources. Appendix D is a simplified identification guide to families and genera that are common in this region. Appendix E is a quick guide to locate where those lichen genera are commonly found in the landscape. A word of caution here: while you *could* look in each of the given microhabitats and find at least one of those lichen genera listed, do not assume that any lichen you find in those microhabitats is one of the genera listed. Remember that this region has a wide diversity of habitats and plant communities and each has common lichens as well as rare lichens that may be specific to a single habitat.

University of Central Oklahoma's Selman Living Lab south of Freedom, Oklahoma, is geologically distinct because of its gypsum outcrops. *Psora russellii,* a soil lichen, becomes encrusted with the gypsum's calcium sulfate crystals with every rain event. The gypsum dissolves in the rainwater and recrystallizes on the lichen surface along with the lichen's own calcium oxalate crystals which it produces as a white powdery deposit called "pruina."

About Lichens

What are they?

Lichens are fungi living symbiotically with algal and/or cyanobacterial cells among its elongated, multi-nucleated cells called hyphae. The fungus is called the "mycobiont." The alga is called the "photobiont." Lichens rarely resemble either algae or fungi. Together, they are a microscopic universe of complex species interactions. There are almost no rules about what lichens are that aren't broken by one or more species.

Where do they grow?

Lichens are found in virtually every type of environment: desert, tundra, forest, prairie, and seacoast. They grow on a variety of substrates exposed to sunlight: trees, rocks, soil, and even attached to rocks under water. Surprising to some is their ability to grow on human-made substrates such as concrete, asphalt roof shingles, and synthetic fabrics.

How resilient are they?

While most lichens look dead when they are dry, they can survive drought, freezing, and flooding. However, fire and smoke from fire are especially damaging to lichen colonies. Some species are more susceptible to air pollution than others and therefore can be used as air quality monitors. Lichens can be collected any time of day or season of the year, regardless of the weather. Except for color, they can maintain their identification characteristics for decades after collection and storage.

How are they identified?

Lichens are classified by their mycobiont. They are identified using taxonomic keys based on combined characteristics of the thallus (vegetative body) type and color; thallus components and surface features; sexual spores and asexual propagules and conidiospores. Common lichen chemicals can be detected by their color reactions to tests using common house-hold solutions such as bleach or potassium hydroxide (which is found in some drain-cleaners), or the reagents can be ordered from chemical supply companies. The presence of hundreds of other chemicals, unique to certain species, can be determined by a process called thin-layer chromatography, or "TLC." However, TLC requires chemical laboratory equipment not available to many beginning lichenologists, so that testing technique is not described here.

Importance of Lichens

The chemical importance of lichens cannot be denied. They produce a variety of acids that aid in eroding even the granitic rock of the Quartz Mountains and Wichita Mountains into soil. They can recycle mineral nutrients in soil, water, and air, as well as organic matter like bark, wood, leaves, and mosses by producing digestive enzymes. When cyanobacteria are present in lichens, their nitrogen-fixing ability contributes to primary succession in new habitats.

Lichens are a staple in the winter diet of elk and other wildlife. Some large lichens were used by Native Americans as food, but more often they have been used for medicines and dyes. The pharmaceutical industry is investigating their possible use in antiviral, antimicrobial, antioxidant, antigenic, anti-inflammatory, and anti-cancer drugs.

Lichens are becoming an important part of the knowledge for solving immunological problems.

Currently, comparisons of historic and current lichen distributions are being used as pollution indicators for the environmental presence of sulfur dioxide (SO_2), hydrogen fluoride (HF), metals, acid rain, oxidants, and organic toxins. There are measurable correlations between the death of some lichens and pollution.

Changes in environmental factors, as simple as temperature and humidity, can alter rates of nitrogen fixation, gas exchange, and enzyme activity. Such variations can cause pigment degradation and differences in lichen morphology.

Recently, the monitoring of change in lichen community composition as related to air quality, climate change, and biodiversity has become an important indicator of forest health. The U.S. Department of Agriculture Forest Service depends on these lichen data for land management.

The interactions of changing environmental conditions complicate the measurements of any one factor. Fortunately, modern pollution monitoring techniques and computer modeling can be used to calibrate the correlation between pollution and morphological changes in lichen populations. These data are becoming more and more important as we experience and adjust our human activities to reduce the effects of pollution and climate change.

There are definitely future lessons that can be learned from lichens. Research is already showing us how DNA from one species can affect the cellular activities of another. Lichens are models of such direct species interactions. Lichens' symbiotic properties hold possible solutions for interspecific immunological problems. However, they are still not noticed in the landscape by the average person, and many people do not even know that lichens exist. There are opportunities to become an experienced lichenologist and contribute to both medical and environmental research.

Finding and Collecting Lichens

To study lichens you will need to observe dried specimens from a herbarium or collect fresh ones from the field. Collection permits are required to collect on state and federal property, and prior permission should be requested of private land owners. To find lichens in the field you may need to do some close observations using a hand lens. It is best to avoid scarring woodlands, rocks, landscape features, or infrastructure in the public view. So only remove lichens from substrates in remote areas and never from cemetery headstones.

Tools for collecting include a hand pruner for clipping lichens on small branches, a sharp knife or wood chisel for shaving lichens from bark, and a geologic hammer, safety glasses, and a rock chisel for collecting rock with lichens on it. The knife or chisel can also be used to scoop lichens from soil. Paper bags are the best containers for transportation and temporary storage of lichens. Plastic bags that inhibit dessication can cause specimen loss due to secondary mold growth.

Observing and Documenting Lichens

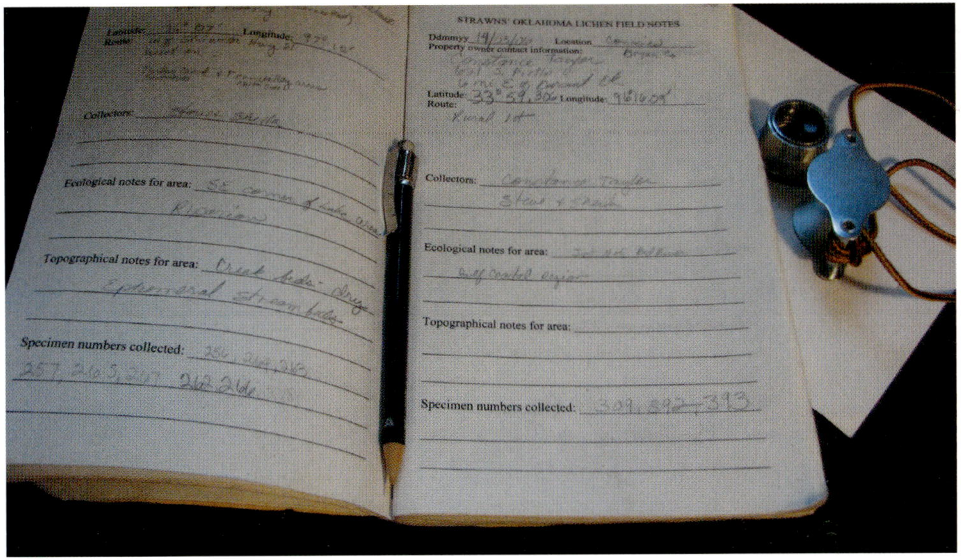

When collecting be sure to record general and specific data such as location, substrate, and habitat descriptions, along with contact information for the land manager. This can be recorded in a field book, but electronic devices are becoming much more helpful since many have applications for GPS location as well as high resolution digital cameras.

If you are collecting in an area covered by internet service, some of the more common lichens can be identified in the field using internet applications. However, a photo taken in the field with a cell phone or camera can also be used in the lab to aid in identifying a specimen by recording colors or characteristics that might have faded or changed after the lichen is stored. More importantly, a digital photo is a valuable record that can be attached to collection data and made globally available.

Identifying Lichens

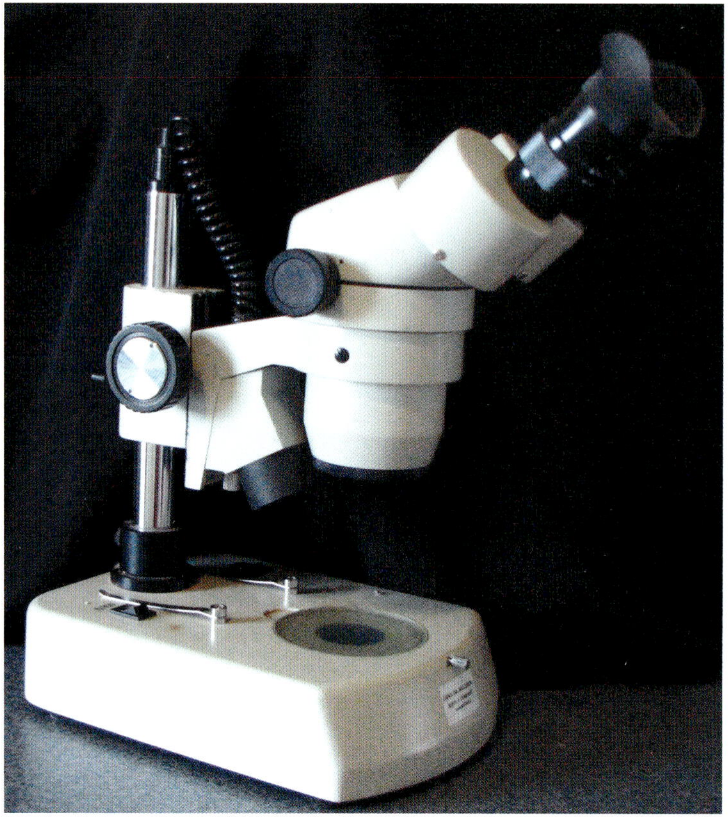

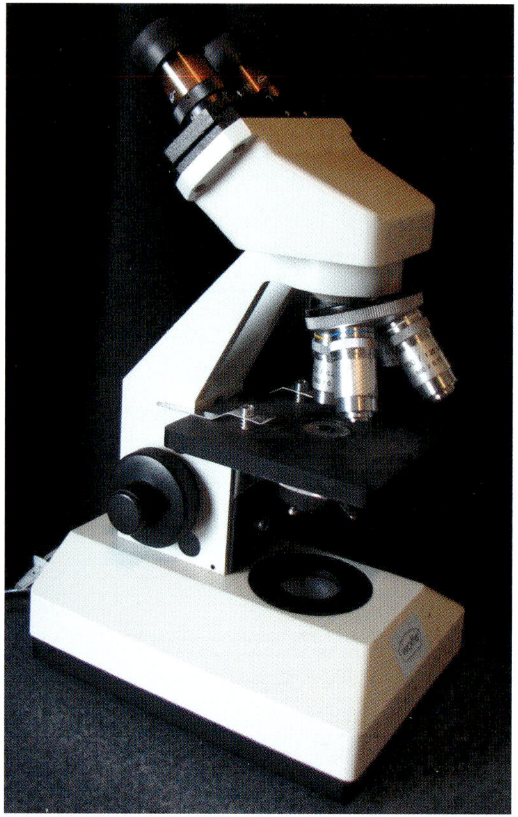

Reliable identification of a specimen usually requires observation of thallus characteristics using a dissecting microscope and of microscopic spore characteristics using a standard light microscope. Use of a lichen key for the specific region in which you have collected the lichen can give you a good preliminary identification, but the chance of correctly identifying the specimen improves when you verify the species using a second key for a larger region.

Before coming to a final determination, compare your specimen to a complete description of a species. It is always possible that you have found a new, undescribed species or one that has not yet been found in this region, or you may simply have missed an important characteristic and be mistaken in its identification. Many keys, species descriptions, and images are available online for comparisons. So if the lichen you are trying to identify doesn't EXACTLY fit the options in the key you are using, use sources listed in the Bibliography or websites listed in Appendix F to find other keys that include more species, as well as additional information. Several sites have links to databases that offer searchable checklists for specific regions.

You can also search the internet to find and connect with individuals or groups of lichenologists. The American Bryological and Lichenological Society (ABLS) and the Mycological Society of America have websites that list activities and events you can attend and they give information about becoming a member. It helps to get personal advice regarding those lichens you have spent many hours trying to identify, and working with other lichenologists can help you develop the skills you need as your knowledge of lichen biology grows.

When documenting specimens it is important to archive as much information about them as you can.

When documenting specimens it is important to archive as much information about them as you can. First, assign each specimen a collection number. For each, it is important to record the collector, date, substrate and habitat description, and exact location or GPS data. In addition, when you have identified the specimen, record the name (genus and species) followed by the name (or abbreviation) of the author who first described the species. The format for these elements and others used in assigning species names to lichens are the same as for botanical collections. In *Caloplaca* sp., for example, "sp." indicates that the species it belongs to has not been determined. If the species has been renamed or transferred to another genus, the original author's name is in parentheses, and the author who placed the species in another genus follows that. Often there have been two or more authors for either or both descriptions; for example, *Caloplaca flavovirescens* (Wulfen) Dalla Torre & Sarnth. was described originally by Wulfen as *Lichen flavovirescens*, and the species was later transferred to *Caloplaca* by Dalla Torre and Sarnth.

Identifying Lichens (continued)

The identification and collection information is attached in the form of a label that is printed or glued on a sheet of 8½ inch × 11 inch archive quality paper, which is folded to hold the specimen. A color image or other information such as location details or a distribution map, can be printed or glued on the outside of the envelope, as well.

Clear, archive-quality acrylic boxes or petri dishes can be used to contain lichens for teaching purposes or to maintain the integrity of lichens on fragile substrates like soil. A box is also a more durable container for lichen specimens on rock. The numbered boxes can be sorted into trays and stored in herbarium cabinets. Their numbered labels can be filed separately or with the specimens.

Preserving and Storing Lichens

Preserving specimens in a herbarium requires freezing the lichens for up to a week before storage. This reduces mold spore contamination and kills insect pests. In southern temperate zone regions like Oklahoma, if not sufficiently treated, insect larvae can hatch from eggs and damage the specimen.

Many herbaria organize lichen specimens by genus and store the folded paper envelopes in metal 5 inch × 7 inch card file drawers or attach the envelopes to herbarium sheets and store them in standard herbarium cabinets.

Thallus Components

A **thallus** is the body of the lichen. Most species have densely packed fungal hyphae in the upper cortex layer with algal cells just below it. A medulla of loose fungal hyphae is just below that. Most lichens have another densely packed fungal layer comprising the lower cortex. However, crustose lichens lack a lower cortex, and the ecorticate leprose lichens consist of only medullary hyphae with algal cells. If there are rhizines, they extend from the lower cortex to the substrate. This cross-section view of thallus structures of *Punctelia graminicola* shows the typical positions of the thallus components. However, each of these components is variable or may be totally absent in some species.

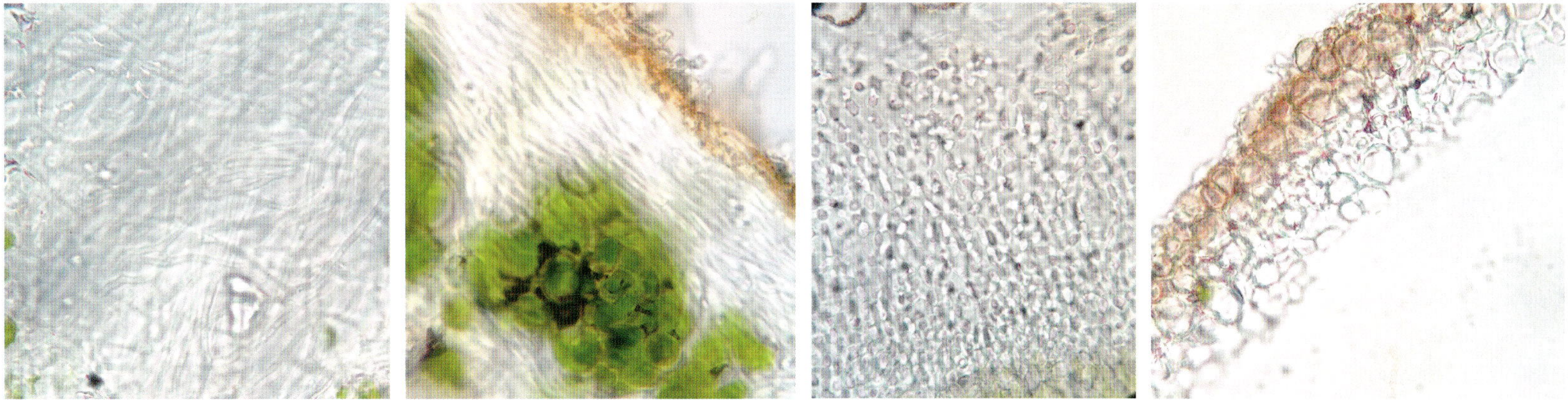

Upper Cortex

The **upper cortex** consists of fungal cells which are usually transparent. Its microscopic texture is characterized by the shape, branching, and orientation of those cells. In many identification keys there are several technical terms used to describe the texture, which can be confusing at first, but are similar to botanical terms. For example, while plants have basic tissues called parenchyma, fungi have groups of similar cells called plectenchyma. When the fungal cells are long hyphae they constitute a "prosoplectenchymal" texture. These hyphae may be loosely interwoven as in *Ramalina* sp. on the left; aligned parallel to the algal layer (periclinal) as in *Teloschistes exilis,* second from left; or aligned perpendicular to the algal layer (anticlinal) as in *Psora* sp., third from left. Otherwise, a lichen's upper cortex may have a "paraplectenchymal" (sometimes called "pseudoparenchymal") texture. Lichens with this characteristic have rows of isodiametrical cells, with the same length and width, as seen on the right in *Dermatocarpon miniatum* (photos by Bob Harris).

Prosoplectenchymal hyphae can be somewhat wavy so that only short segments of the filaments are in focus under the microscope. This causes them to appear to be rows of slightly elongated cells. Also, if they are cut transversely across their diameter, they can appear to be isodiametrical, like paraplectenchymal hyphae. So this characteristic may be very difficult to discern. If your identification key uses these terms and you are unable to determine which texture your specimen has, try using other characteristics to identify your specimen, or use a different key.

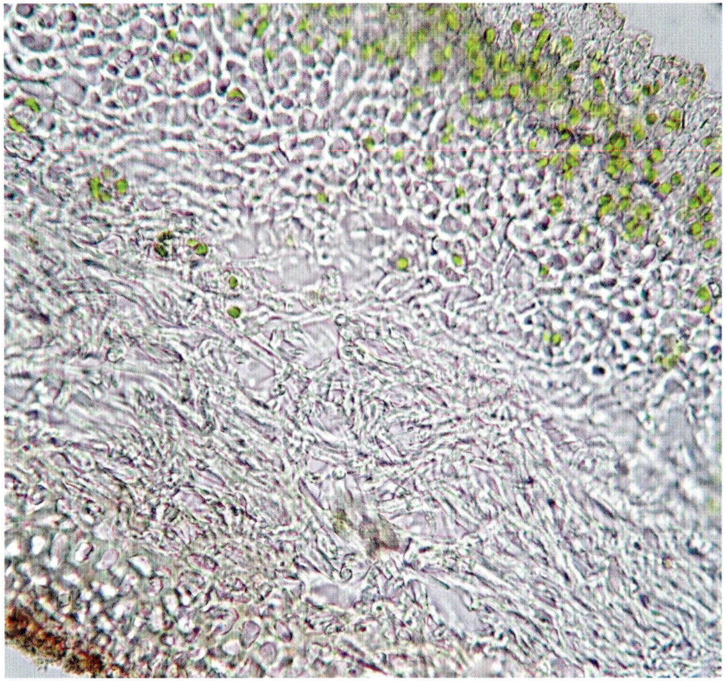

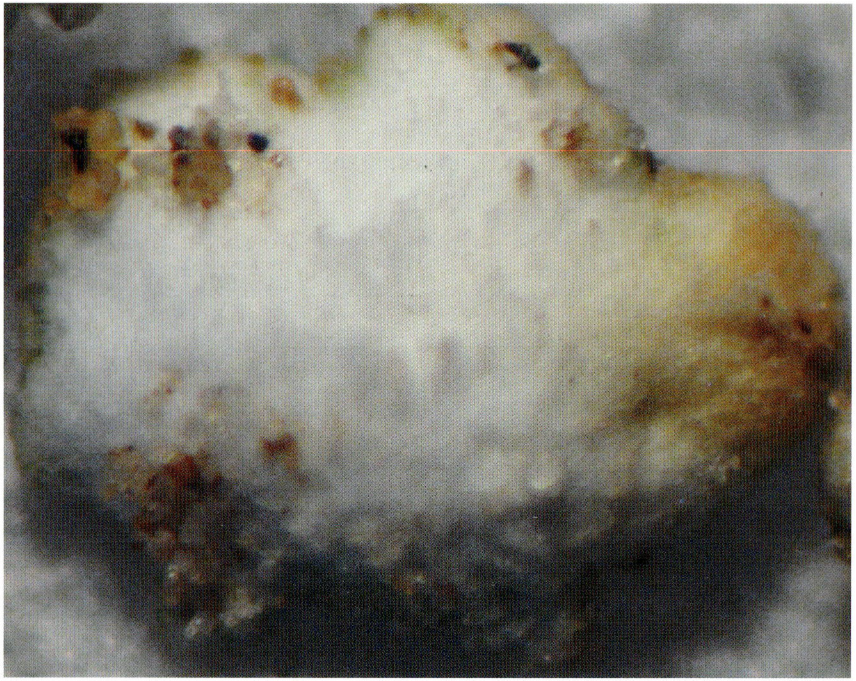

Medulla

The **medulla** is the middle layer of loosely woven fungal hyphae between the upper and lower cortexes as shown above left in *Dermatocarpon miniatum* (photo by Bob Harris). Often the thickest layer, the medulla serves as an air reservoir for the green algal or cyanobacterial photobiont. Medullary hyphae are hydrophobic, so water transfers directly from the medullary layer into the photobiont cells above it.

Primary squamules of *Cladonia* species lack a lower cortex, exposing the medulla. Medullary hyphae are usually white as seen on the right in *Cladonia peziziformis*, but hyphae of some species produce pink, orange, or yellow pigments.

Lower Cortex

The **lower cortex** is another layer of either prosoplectenchymal or paraplectenchymal fungal cells. It may be wrinkled or smooth and be white to brown or black. It may have a tomentum of fuzzy, colorless hyphae or be attached to the substrate by rhizines made of prosoplectenchymal hyphae. Rhizines may be black or white and may be unbranched as seen in *Parmotrema conferendum* on the left, or branched as seen on *Punctelia graminicola* on the right.

The lower cortex of some fruticose, foliose, and squamulose lichens that lack rhizines may have a variety of specialized attachment structures such as an umbilicus, which is a single bundle of prosoplectenochymal hyphae or by several hapters, or holdfasts, which are also bundles of such hyphae. Crustose lichens, which lack the lower cortex, may be directly attached to the substrate by medullary hyphae.

Thallus Types

Foliose lichens may have a narrowly branching, leaf-like thallus such as that of *Xanthomendoza weberi*, left. Other foliose lichens may have rounded lobes, like *Parmotrema reticulatum*, right.

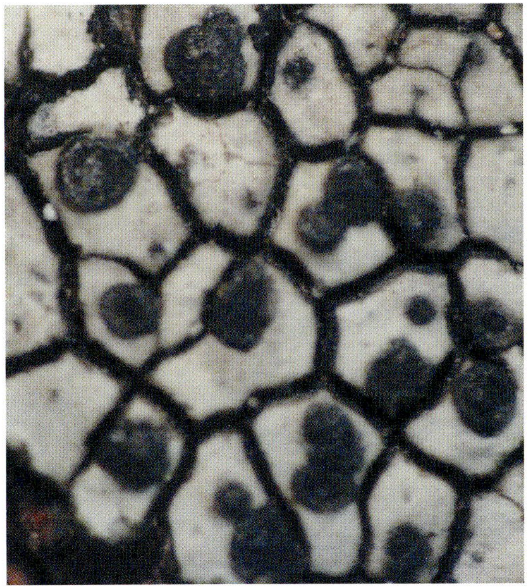

Crustose lichens may have a continuous thallus that is so tightly appressed to the substrate that it appears to be paint on rock or bark, like *Arthonia susa* (above left).

Buellia spuria (center) has an areolate crustose thallus which appears to be broken into discrete patches or tiles.

Dimelaena oreina (right) is an "effigurate" crustose lichen, showing a transition between crustose and foliose growth forms, with lobed areoles on the thallus margin.

Thallus Types (continued)

Leprose lichens are a special type of crustose lichen. Their sterile continuous thallus has no upper or lower cortex as seen on the left in *Lepraria finkii.* They consist almost entirely of soredia. The soredia are algal cells within clusters of fungal hyphae. They may appear to be a coating of light green powder or fuzzy balls as shown at right in the closeup photo.

Endolithic lichens (not shown) are another special type of crustose lichen. They lack the tiny balls of soredia and are simply algal cells with fungal hyphae growing several millimeters into the porous or transluscent surface of rock such as limestone or course-grained granite, giving the appearance of green stain rather than paint or only showing their ascomata.

Squamulose lichens sometimes appear to be foliose but are more scale-like, rather than leaf-like. The appressed or "adnate" squamules of *Placidium lacinulatum* (left) lie horizontally on the substrate.

However, many *Cladonia* species like *Cladonia polycarpoides* (right) have basal "primary" squamules that are attached on one edge and are oriented perpendicular to the substrate.

Thallus Types (continued)

Fruticose lichens have a three-dimensional growth pattern, so it can be difficult to distinguish between their upper and lower cortexes. *Ramalina sinensis* (left) is an erect fruticose lichen that is stiff enough to allow its short 2 to 3 inch (5–8 cm) thallus to stand upright from the substrate.

Cladonia subtenuis (right), which can grow up to 6 inches (9 cm) in height, is stiff due to an inner cartilaginous medulla which keeps its intense branching pattern raised above the forest floor, allowing light to reach the lower branches of the lichen. This species, like many other *Cladonia* species, has no basal squamules, only highly branched podetia that develop from a temporary crustose thallus.

Usnea dasaea (left) has a flexible fruticose thallus that hangs down approximately 5 inches (12.5 cm) from a single base. This pendant type of thallus is more common in moist areas such as Robbers Cave in southeastern Oklahoma.

Several *Parmotrema* species (right) appear fruticose, but are referred to as "ascending foliose" rather than fruticose.

Thallus Features

Cilia are special rhizines that are usually longer and may be thicker than rhizines that extend from the lower cortex. Some are white as on the reproductive cups of *Heterodermia echinata* at left, but many are black and project from the edge of the thallus as in *Parmotrema hypotropum* at right.

Isidia are perpendicular growths on the upper surface of the lichen that contain medullary hyphae covered by the dense upper cortical hyphae with algae as in *Myelochroa obsessa* (left). They function mainly to increase surface area, but they are brittle and can break off, propagating new lichens.

If isidia-like growths appear to be flattened as in *Punctelia graminicola* (right), they may be called lobules.

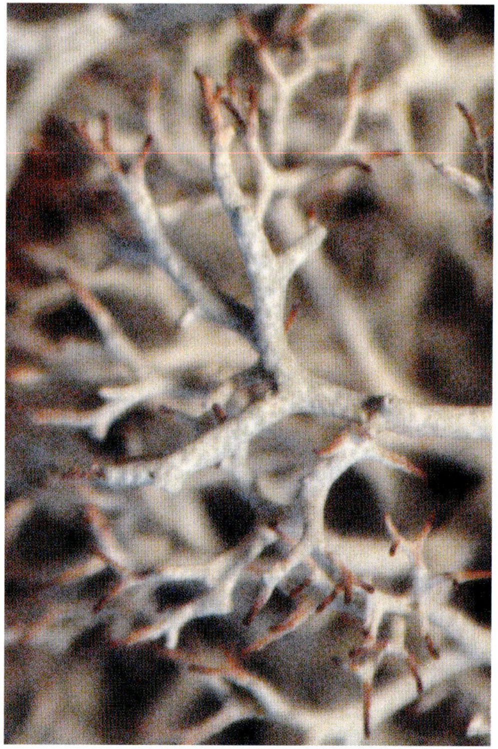

Thallus Features (continued)

Podetia are stalk-like structures that usually support sexual reproductive structures in many species. Here, terminally branched podetia support red apothecia in *Cladonia cristatella* on the left. However, intricately branched podetia of *Cladonia subtenuis* (right) lack cups or caps.

Pruina is a powdery or grainy covering of the thallus that is white to gray. It consists of dead cells or calcium oxalate crystals formed by the precipitation of calcium absorbed from the substrate, or of other materials. Here you see pruina on the upper cortex of the thallus of *Pyxine subcinerea* at left.

Pseudocyphellae are holes in the upper cortex of the thallus that expose the medullary fungal hyphae. Such a feature on the lichen surface is described as "punctate." Pseudocyphellae occur in all *Punctelia* species such as *P. bolliana* shown here at right. Incidentally, "cyphellae," which are recessed breaks in the lower cortex, are very seldom seen in lichens in this region. Neither of these should be confused with "cephalodia," which are dark blue-green, gall-like growths containing cyanobacteria in the upper or lower surface of a thallus.

Thallus Features (continued)

Pycnidia are small black dots on the thallus surface that contain asexual reproductive fungal spores called "conidia." They are most often near lobe margins as they are in *Punctelia bolliana* (left).

Soralia are where soredia (clusters of asexual medullary fungal and algal cells) are released through openings in the upper cortex of the thallus. Apothecial cup margins of some species can also have soredia as shown at right in *Parmotrema reticulatum.*

Pigments and Coloration

A **gray** thallus like this *Physcia aipolia* (left) is the most common lichen color. This color is usually due to atranorin, a pigment in the upper surface. When it is wet, the lichen may appear greener because the upper cortex becomes more transparent, allowing the green algae to show through. Under different lighting conditions an individual species can vary in color from gray to green to yellow-green, depending on the production of secondary chemicals which are not directly responsible for metabolism.

Sometimes similar species of lichens with different pigments can grow into each other and give the appearance of a single bi-color lichen as in this photo (right) of gray-green *Punctelia rudecta* and yellow-green *Flavoparmelia caperata* which produces usnic acid, a yellow pigment. An individual species is not likely to have such color variation on the same substrate.

Pigment and Coloration (continued)

Several secondary pigments provide protection from solar ultraviolet (UV) light or serve as a deterrent to predation. Anthraquinone pigments produce an **orange** color in *Caloplaca subsoluta* (left) and in most other species in the Teloschistaceae family.

Rhizocarpic acid, a derivative of pulvinic acid, gives a **yellow** to **chartreuse** color to the crustose lichen *Acarospora novomexicana* (center). The pigment also fluoresces bright orange when tested under long-wave UV.

The **black** thallus of *Lichinella nigritella* (right) is due to the presence of its cyanobacterial photobiont, rather than a secondary pigment. The photobiont in the lichen turns blue-green when it is wet and exposed to light.

Psora decipiens (left) has black apothecia around the margins of its thallus squamules which are often **brick-red** to **orange-brown,** but the thallus can appear pink on the gypsum soils of Selman Living Laboratory south of Freedom, Oklahoma.

Diplotomma venustum (right) is **white** due to the pruina covering its surface. However, over time black hyphae of parasitic fungi can spread through the older parts of the thallus, giving it a grayish appearance.

What Color is a Dead Lichen?

Lichens grow continuously and may be collected any time of day or season of the year. Most can survive temporary drought, freezing, or flooding, but not fire. Understandably, lichens on some substrates are more susceptible to fire than others. Some cannot tolerate even temporary or slightly altered conditions like smoke or increased exposure to sunlight. Furthermore, many cannot even survive being moved to environmentally controlled conditions of a laboratory growth chamber. Determining when a lichen is dead or dying can complicate research or species inventories you might be conducting.

After both the algal and fungal cells die, altered nutrient supplies, secondary xanthones or anthroquinone pigments may linger, giving the thallus an orange or pinkish color. Lichens like *Parmotrema perforatum* that contain salazinic or norstictic acid can become red as they deteriorate.

When only the photosynthetic alga dies, the lichen's fungal cells are left intact until the fungus eventually deteriorates from lack of nutrients as in this specimen of *Parmotrema perforatum*. However, don't assume that a white lichen is dead or dying. You can scrape the surface to see if white calcium oxalate crystals are covering its upper cortex. You can also drop water on it and hold it in the light. The upper cortex will become more transparent, and if its algal photobiont is still alive its chlorophyll will show through, causing the lichen to appear greener. If it doesn't turn green the specimen could be dead, or it could be a non-lichenized fungus.

Sexual Reproductive Structures

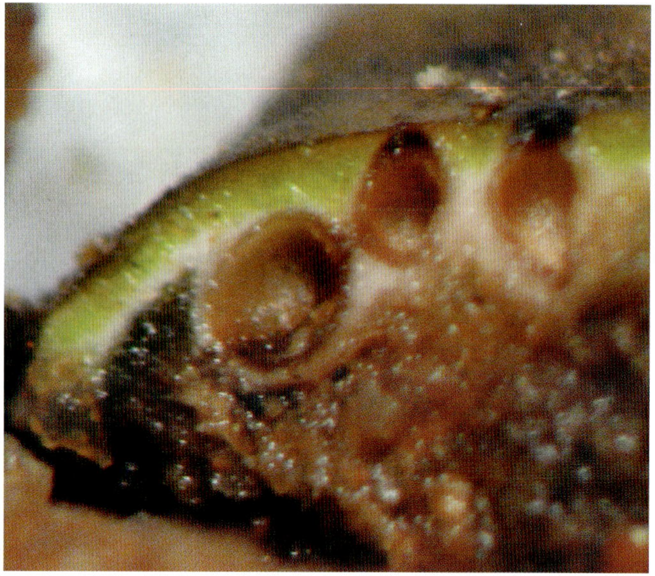

Based on their sexual reproductive structures, lichens may be classified in either of the fungal phyla Basidiomycetes or Ascomycetes. Basidiomycete lichens, which produce spores externally on microscopic stalks, are rare in this region and are not described here. The Ascomycete lichens, which produce sexual spores in microscopic sacs called "asci," are by far the most common. Species of *Lepraria* which have no sexual reproductive structures have also been placed in Ascomycetes. In fertile Ascomycetes the structures that hold the asci are called ascomata. There is a variety of ascomata shapes based on their outer appearance. Whether they are in a pit or lie on the surface as a disk, cup, or groove, they are usually on or in the upper cortex of the thallus.

Types of Ascomata

Perithecia are ascomata that are like flasks buried in a pit in the thallus, showing only a narrow opening of the ascomata on the surface. That small opening is called the "ostiole." The empty perithecial cavities in the cross-section of this *Placidium squamulosum* show that the ostiole develops with maturity, finally erupting on the surface as seen on the right.

Apothecia lie on or above the thallus surface. They hold the asci in the upper surface of a disk, cup, or groove. That upper surface is called the "hymenium." It is exposed as the spores develop as seen in *Physcia aipolia* and its cross-section.

Types of Ascomata (continued)

"Lecanorine" apothecia may be cup-shaped like *Punctelia bolliana* (left) or disk-shaped with the thallus growing like a rim around the disk, like *Candelaria fibrosa* (right).

If the disk's rim is not the same color as the thallus (the thallus is not wrapped around the apothecial disks) and the disks are black, they are "lecideine" apothecia as is *Lecidella enteroleucella* (left). If they lack the thallus rim but the disks are not "carbonized" (black), they are called "biatorine." However, except for some *Psora* sp., lichens with biatorine apothecia are uncommon in Oklahoma.

Apothecia in the shape of elongate grooves in the upper cortex are called "lirellae." While most lirellae are superficial, some are buried in the thallus, somewhat similar to perithecia. Lirellae of lichens, such as *Graphis scripta* (right), give the appearance of being hand-written script on bark.

Types of Ascomata (continued)

On fruticose lichens, apothecia may be flat and terminal (at the tips) or anywhere along the branches as in *Teloschistes exilis* (left).

Some squamulose lichens have apothecia on the ends of secondary fruticose thalli called "podetia." These apothecia are sometimes so convex that they appear to be caps as in *Cladonia peziziformis* (right).

If the tips of the podetia are flared, tiny apothecia or pycnidia may ring the rims of the cups as seen here in *Cladonia grayi*.

Asci and Spores

Spore characteristics are important for the identification of many lichens but are impossible to determine without a microsccope. Even with a microscope, identification can be difficult without knowing how to mount a cross-section of the spores on a slide using a common wet mount technique.

Cross sections of apothecia or perithecia can be made using a very sharp razor blade as shown here in the photo of *Physcia biziana*. Cutting several very thin sections increases the chances that you will get a slice transparent enough to allow light to expose the spores within the asci inside the apothecium. To view the sectioned apothecium under the microscope, immerse the separated sections in a drop of water and apply a coverslip.

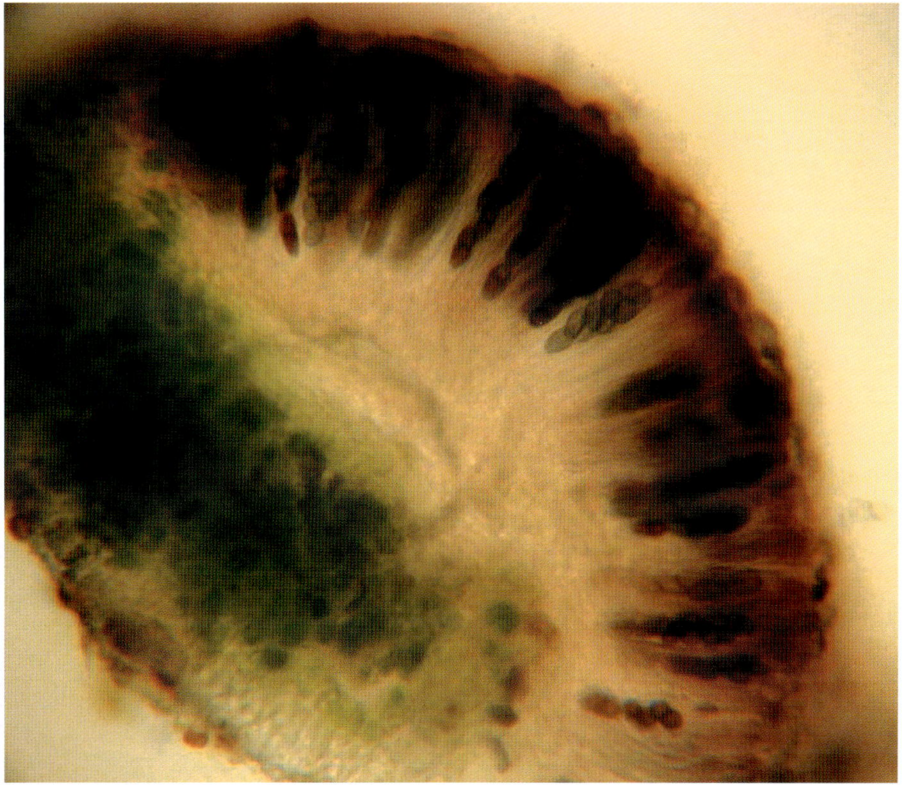

Asci, the spore sacs in the ascomata, may be in the pit of a perithecium, in the groove of a lirella, lining an apothecial cup, or in the upper surface of an apothecial disk.

This photo of a cross-section of a disk was taken through a light microscope at 400× power. It shows *Hyperphyscia syncolla* in which the asci are packed vertically between sterile hyphal filaments called "paraphyses," with eight 2-celled spores per ascus. In some species there are no paraphyses, and in some species spores are "mazedial" (dry and loose) on the open surface of the apothecium.

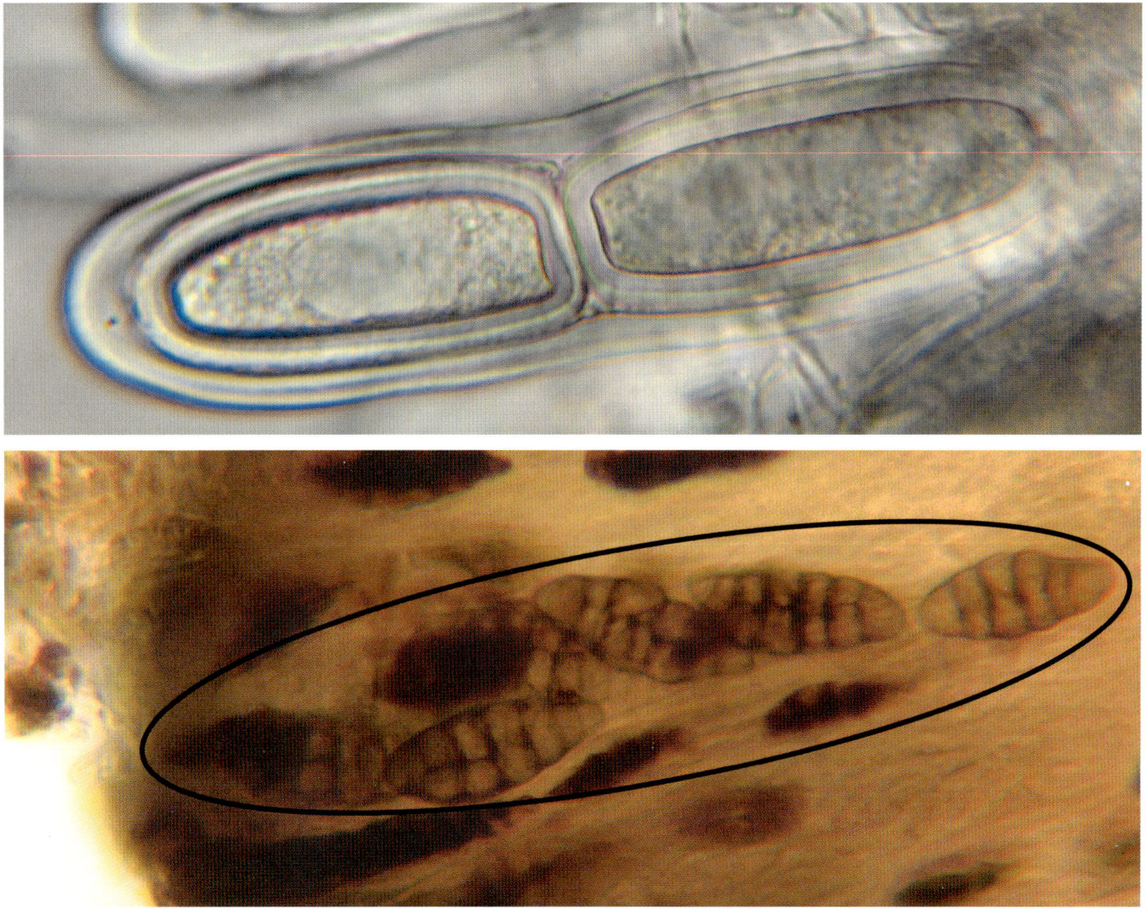

Asci and Spores (continued)

Pertusaria pustulata (top) has two large single-celled spores per ascus. The black line approximates the ascus wall in *Diploschistes diacapsis* (bottom). It has eight muriform (having both transverse and longitudinal septa) spores per ascus. Septa are the walls between cells in multi-celled spores.

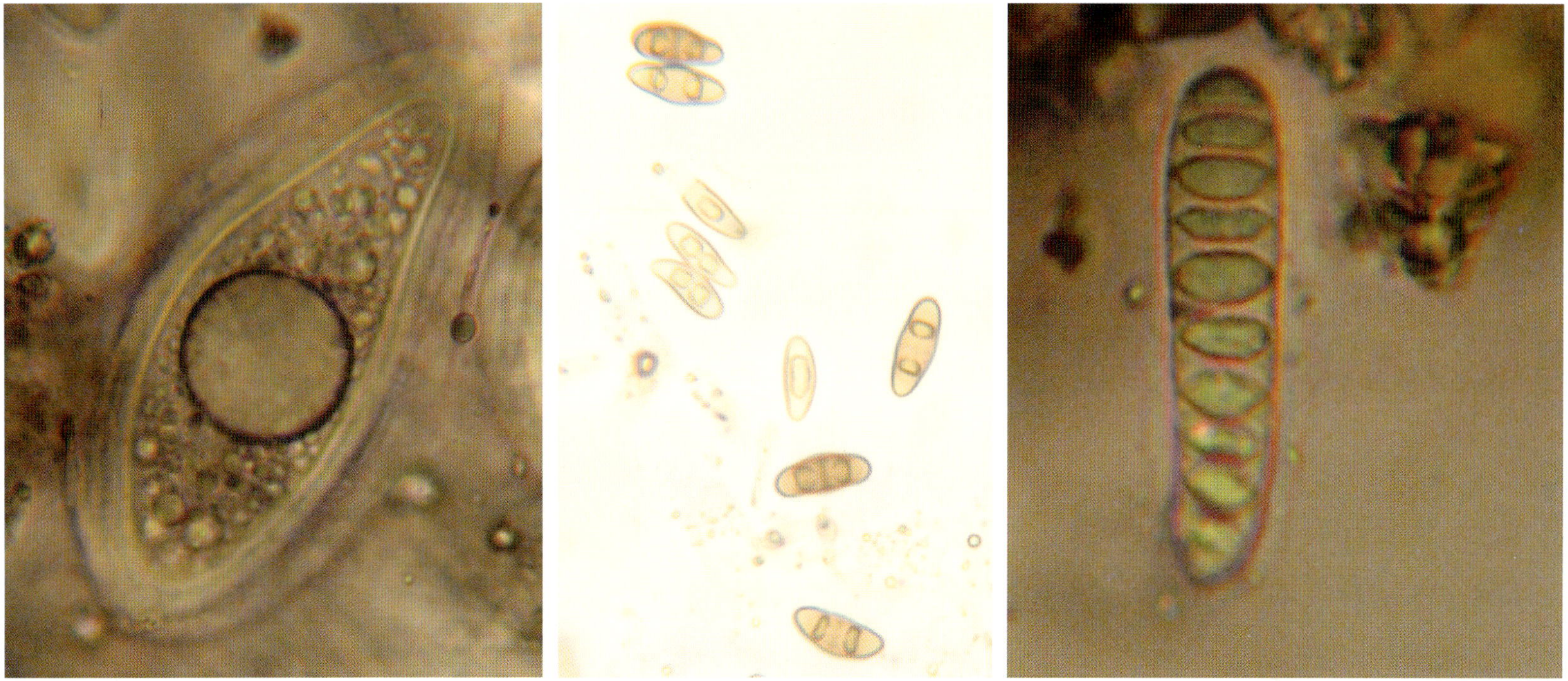

Spores in the asci may be single- or multi-celled. On the left you see the single-celled, eliptical spore of *Pertusaria leioplaca*. In the middle you see the two-celled spores of *Physciella chloantha*. *Graphis scripta* on the right has multi-celled spores that are transversely septate.

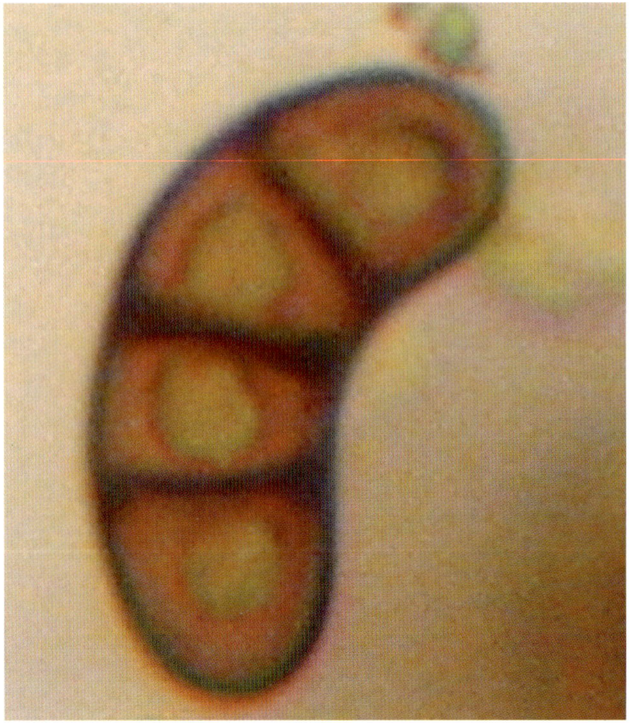

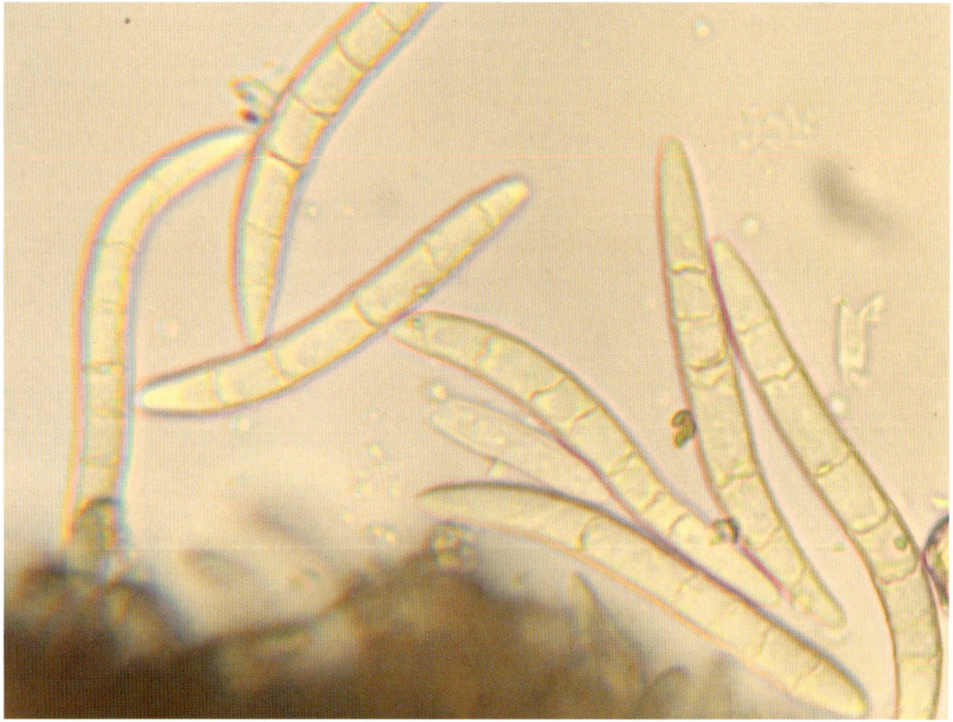

Asci and Spores (continued)

Buellia alboatra (left) has 4-celled (3-septate) spores that are about 15–20 × 8–10 µm. *Candelariella spraguei* (right) has curved to S-shaped cylindrical spores that are 34–48 × 3 µm with several septa.

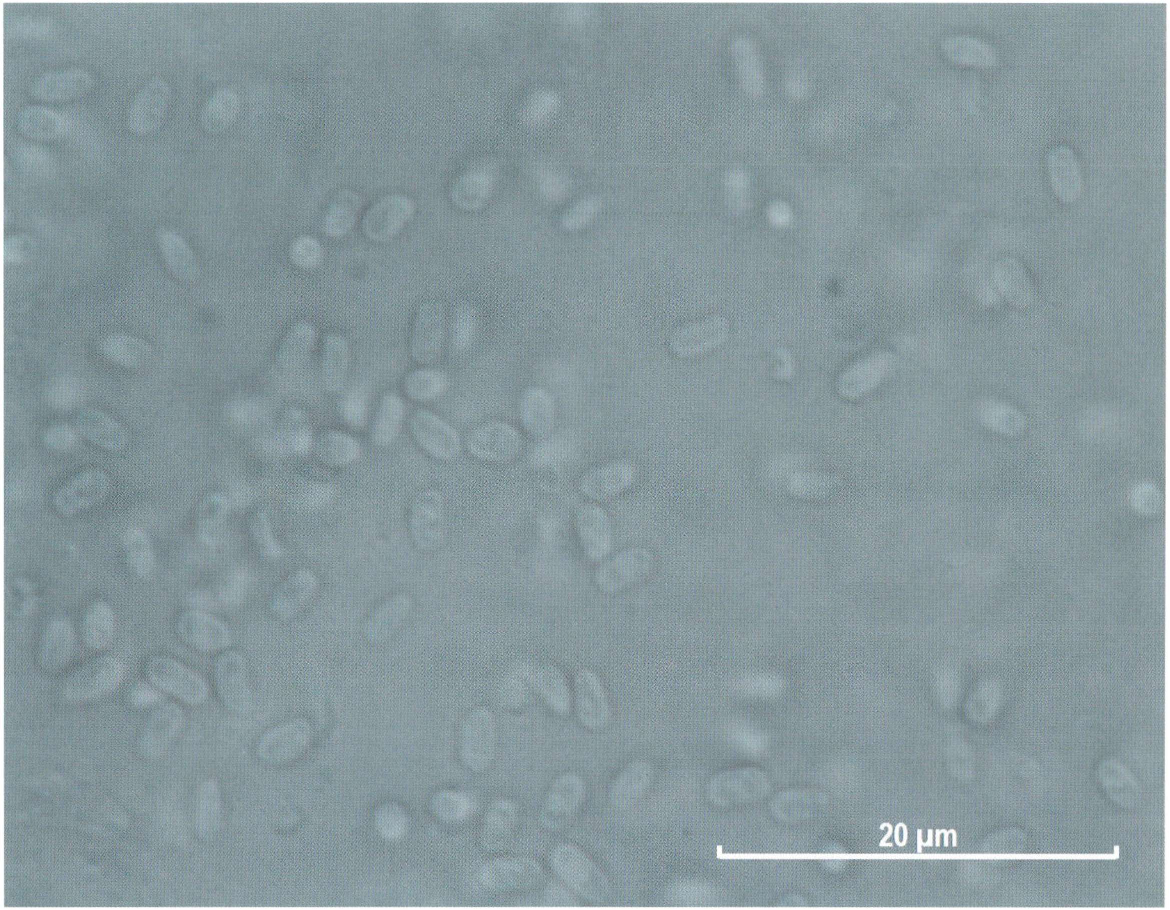

Like other fungal species, lichens can produce either or both sexual and asexual spores. **Conidia** are the asexual spores produced in pycnidia. They are generaly much smaller than sexual spores formed in ascomata as you can see from the scale in this photo of conidia from *Placidium squamulosum* that are approximately 2.5 µm in length (photo by Caleb Morse).

Substrates Where Lichens Grow

Terricolous lichens live on damp soil. Surprisingly, even though western Oklahoma's climate is relatively dry, its clay and gypsum soils can hold water for long periods of time.

Crustose "jelly lichens" and "rock olive" lichens that have cyanobacterial photobionts such as this *Peltula patellata* (left) can grow on red clay which periodically dries out. In the southeastern part of the state, soils receive greater rainfall and maintain sufficient moisture to support large fruticose lichens such as *Cladonia subtenuis* (right) which grows on woodland soils.

Lignicolous lichens grow on bare wood, while **corticolous** lichens grow on bark. Many **saxicolous** lichens, which grow on rock, can produce organic acids and other substances that contribute to the breakdown of their substrate.

Punctelia missouriensis (left) can live on either bark or rock. *Pleopsidium flavum* (right) is a striking chartreuse lichen with an effigurate (appearing to be lobed) margin, which grows on sheer granite surfaces.

Substrates Where Lichens Grow (continued)

Siliceous granite and sandstone, calciferous limestone and gypsum, and basalt all have different chemistries and water-holding capacities that can determine which lichens can grow on them.

Here, the fruticose lichen *Usnea dasaea*, as well as a variety of foliose lichens, grows on the sheer sides of porous sandstone in southeastern Oklahoma. Some lichens get nutrients or minerals from their substrates, others do not. While rock itself contributes no organic molecules, calcium salts from limestone or nitrogen from bird droppings can be used by some lichens.

Epiphytic lichens grow on plants. Magnolia tree leaves and pine cones of several species of evergreens remain on the tree two or more years. This is sufficient time for some of the relatively faster-growing lichens to become established before the leaf or cone falls from the tree and its nutrients are recycled. Here, *Lecanora strobilina* (left) grows on a pine cone bract.

Muscicolous lichens, including many *Cladonia* species, live on mosses, and **lichenicolous** lichens live on other lichens, sometimes parasitically. *Diploschistes muscorum* (right) can completely overgrow *Cladonia* colonies which are growing on moss.

Substrates Where Lichens Grow (continued)

Human-made substrates can support lichens, too. Since lichens need only sunlight and water, they are frequently found on substrates like paint, asphalt, and concrete. For example, the light gray *Physcia aipolia* at left is growing with several other foliose species on a painted metal fence post at the Tallgrass Prairie Preserve northwest of Tulsa.

Almost anywhere in the region, asphalt roof shingles can provide the yellow *Candelaria concolor* (right) a protected place in the sun.

Parmotrema praesorediosum (left) is able to grow on aged, treated timbers, while *Xanthomendoza weberi* (right) grows on sandstone buildings and *Caloplaca feracissima* grows on the building's cement foundations.

Chemical and UV Tests for Identification

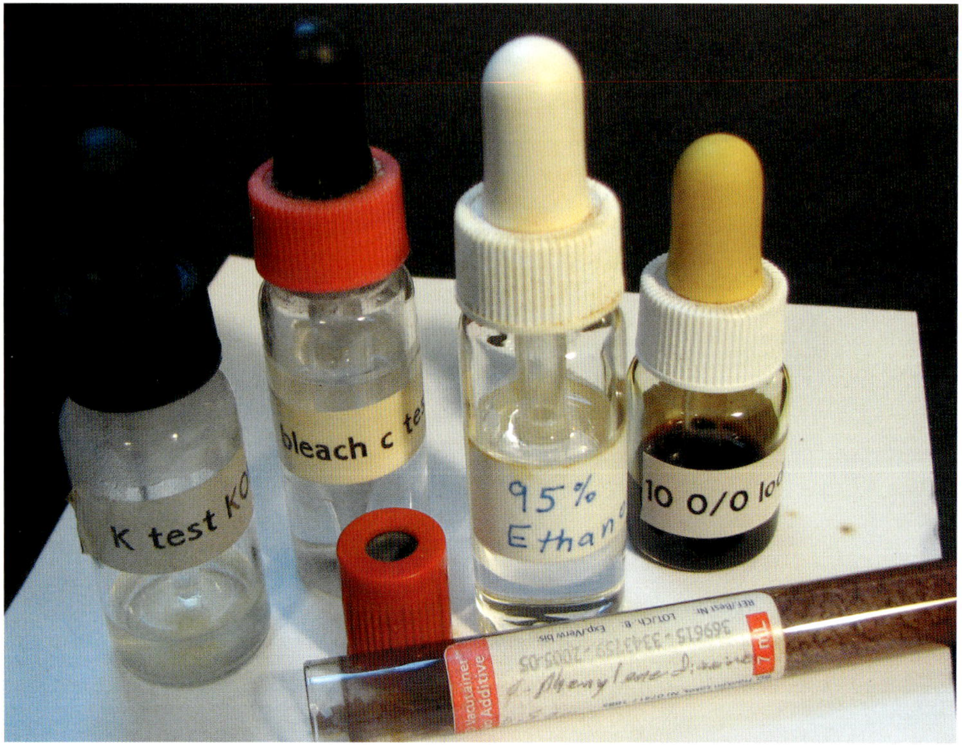

The ability to produce secondary chemical compounds, those not used for photosynthesis, can improve the viability of lichens in several ways. Some pigments protect the lichen from cell damage due to ultraviolet radiation. There are others that may also protect the lichen from parasitic fungi, bacteria, or predation by insects. Some chemicals make fungal cell membranes more permeable, improving their ability to absorb carbohydrates from the algal photobiont. After color, shape, and reproductive structures, much of lichen identification rests on testing for the presence of those secondary chemicals using long-wave UV light and a variety of chemical reactions. However, some chemical color changes are extremely subtle. It takes good color perception and experience to be confident of some species' chemical reactions.

 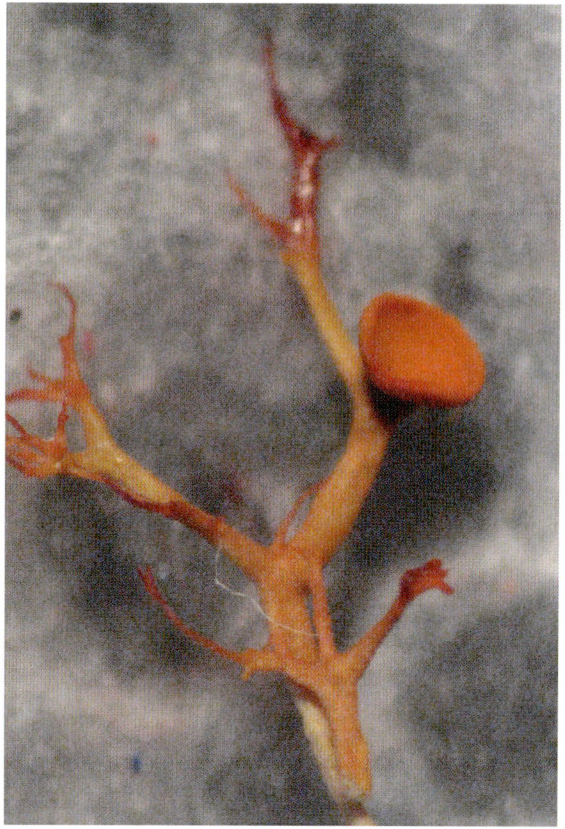

Whether the chemical test is done on the medulla or cortex, the solution can be applied with a dropper or with a toothpick wetted with the test solution. However, when the medulla is to be tested, the cortex must first be scraped off of it as shown on *Punctelia rudecta* (left).

You may need to test several areas and watch the reaction under a dissecting microscope because chemical reactions can be spotty as this *Teloschistes exilis* (right) shows in this K+ red (to deep purple) reaction. Also, watch the reactions for at least one full minute because the colors can change from yellow to orange, red, or purple, or can fade completely.

Chemical and UV Tests (continued)

C tests can be done by applying a drop of either calcium hypochlorite or regular household chlorine bleach. A C+ reaction turns red for gyrophoric acid in *Ochrolechia* species and lecanoric acid in *Punctelia* species. Observing the reaction under magnification as it is being done improves the validity of this test. The scraped surface of *Punctelia rudecta* is shown before C test (left) and after the C+ red reaction (right).

K tests are done with a 10–15% solution of potassium hydroxide. A K+ reaction turns a cortex or medulla yellow or red due to alectorialic, stictic, norstictic, and salazinic acids, as well as atranorin. K+ tests also turn *Caloplaca*, *Teloschistes*, and *Xanthoria* species' thalli to deep purple for the pigment parietin. Potassium hydroxide in the K solution often changes the appearance of the lichen being tested. The same thallus cortex and apothecia of the yellow lichen, *Caloplaca subsoluta*, are shown here before (left) and after a K test (right).

KC tests done on species from Oklahoma have seldom given definitive results different from what the K or C tests give individually. Furthermore, the color change is very subtle and sometimes difficult to detect. It requires sequential application of K and C. A KC+ reaction turns gold in the presence of usnic acid, but turns red in reactions with alectorialic acid. If you perform a KC test, use the same application method as for K or C tests. Drop K solution first and wait for it to be absorbed, then drop C solution onto the sample within a few seconds. When some identification keys call for a CK test, simply reverse the order.

Pd tests are commonly done with a few phenylenediamine crystals dissolved in 95% ethanol, but it can be done just as effectively using 70% isopropyl alcohol. A positive Pd test turns the cortex or medulla yellow, orange, or red due to the various aromatic aldehydes that some lichens produce. The simplest way to do this is to use tweezers or a tiny spatula to lay one or more crystals on the lichen to be tested and to drop a single drop of alcohol on it.

Here you see *Cladonia peziziformis*, first with the pink Pd crystal lying on one of the squamules (left), then the positive red reaction after the application of alcohol (right). After the color change is noted, the remaining crystals can be discarded in a safe manner. Be careful handling Pd because it stains clothes, paper, flooring, etc. Pd is also a carcinogen, so discarding more than a few tiny crystals at a time constitutes a biohazard. Making a stock solution that quickly outdates is not recommended, although several formulas exist that can be stored as long as a couple of weeks. Disposing of an outdated stock solution is a biohazard. Pd powder and crystals have a longer shelf life when kept dry and away from alcohol and sunlight, and they can be safely disposed of, a few crystals at a time, as you use them. If you have trouble obtaining phenylenediamine, try collaborating with a local college laboratory.

UV fluorescence is tested using a long-wave (approximately 360 nm) black light. A UV+ reaction is caused by lichen chemicals that fluoresce specific colors in the dark under black light. Warning: do not look into the lamp when it is on. The thallus of *Acarospora contigua* appears yellow-green under natural light (left) and UV + orange under ultraviolet light (right).

Appendix A: Supplemental activities

1. Access several of the lichen websites in Appendix F or other websites and write summaries of the information available on those sites, suggesting when a lichenologist would need that information.

2. Design an experiment by which you can determine which substrates can hold the most water. Then carry it out and write a report or make a multimedia presentation you can give to other lichen enthusiasts.

3. Do an internet search to find three papers that reported research on the effects of pollution on lichen species. Choose the one that has the most convincing data and briefly describe the methods used. Be sure to provide proper bibliographic citations. If methods and materials are available, carry out that pollution monitoring exercise on a natural lichen habitat.

4. Use the "Thallus Characters Key" in Appendix B to identify specimens of each thallus type in the key.

5. After observing 10–20 specimens in a single family of lichens, make a list of species in that family that have each of these thallus features: cilia, isidia, pruina, podetia, pseudocephellae, pycnidia, and soralia.

6. Cross section a fruticose or squamulose thallus under the dissecting scope. As you examine it, sketch it or photograph and print it out. Label the upper and lower cortexes and the medulla as well as other thallus features that are visible.

7. Make a very thin cross-section of a thallus and examine it under 1000× magnification. Draw or make a digital image of the shape, orientation, and branching of the fungal cells in its upper cortex. Determine whether its texture is paraplectenchymous or prosoplectenchymous and periclinal or anticlinal.

8. Make thin cross-sections of an apothecium. As you examine it under 400× magnification, draw or make a digital image of the asci with one or more individual spores. Label the asci, paraphyses, spores, and hymenium.

9. Use the "Chemistry Test Key" in Appendix B to identify at least 3 lichens in the key, using chemical spot tests.

10. Research and write a procedure for preparing K testing solution. Test it on known K+ and K-specimens.

11. Research methods of pigment extraction from lichens. Identify the pigment responsible for a particular color and collect the species which produces it. Dye a swatch of white cotton fabric or wool yarn with one of the methods you find.

12. Prepare saturated aqueous solutions of calcium oxalate (CaC_2O_4) from a *Pyxine* thallus, calcium sulfate ($CaSO_4$) from gypsum crust soils, and sodium chloride (NaCl) in common table salt. Then precipitate crystals from a drop of each solution on a glass slide by drying it under a lamp. Observe them under the microscope and draw or make a digital image comparing the different shapes and sizes of crystals. Measure and describe the differences.

13. Collect several lichen specimens and prepare them for deposition in a herbarium. Prepare labels with all necessary collection information.

14. Research lichen DNA extraction techniques and find websites for laboratories that will perform DNA analysis. Then prepare and send a specimen for analysis according to their protocol.

Appendix B: Practice keys for lichen identification

Each is specific for the specimens named in the key.

Practice key based on thallus characters

1. Foliose or fruticose.
   2. Foliose.
      3. Thallus black umbilicate, on limestone, wart-like apothecia, no soredia ________ **Lichinella nigritella**
      3. Thallus gray to green to yellow thallus.
         4. Lobate, punctate, apothecia, no soredia ________ **Punctelia bolliana**
         4. Lobate, not punctate.
            5. Soredia marginal or terminal ________ **Parmotrema reticulatum**
            5. Soredia absent ________ **Xanthoparmelia lineola**
   2. Fruticose.
      6. Thallus green, branches flat, apothecia on strap-like thallus ________ **Ramalina sinensis**
      6. Thallus yellow, branches terete, apothecia on bush-like thallus ________ **Teloschistes exilis**
1. Squamulose or crustose.
   7. Squamulose.
      8. Thallus green with brown apothecia on podetia ________ **Cladonia pezizformis**
      8. Thallus green to yellow-green with brown to black apothecia, not on podetia ________ **Psora icterica**
   7. Crustose
      9. Thallus yellow, on calcareous (limestone) rock, orange apothecia immersed in thallus ________ **Caloplaca flavovirescens**
      9. Thallus yellow, on siliceous (granite or sandstone) rocks, brown apothecia immersed in thallus, >1 per areole ________ **Acarospora contigua**

Practice key based on chemical tests

1. Foliose or fruticose.
   2. Foliose.
      3. Thallus green, medulla K+ yellow __________ **Physcia aipolia**
      3. Thallus green, medulla K–.
         4. Thallus green, medulla K–, pseudocephellate __________ **Punctelia bolliana**
         4. Thallus green, medulla K–, not pseudocephellate __________ **Physcia biziana**
   2. Fruticose.
      5. Thallus green, K–, P+ red, UV–, on soil or moss __________ **Cladonia subtenuis**
      5. Thallus yellow, K+ red or K–, P+ yellow, on bark __________ **Usnea strigosa**
1. Squamulose or crustose.
   6. Crustose.
      7. Thallus white, UV–, medulla K+ yellow, P– or P+ yellow __________ **Buellia spuria**
      7. Thallus yellow.
         8. Medulla and apothecial margin K+ red-purple, P– __________ **Caloplaca flavovirescens**
         8. Medulla and cortex K–, P–, UV+ orange.
            9. Center areolate, with marginal lobes __________ **Pleopsidium flavuum**
            9. Thallus areolate, with no marginal lobes __________ **Acarospora contigua**
   6. Squamulose.
      10. Thallus brown, cortex and medulla K–, P–, C–, squamules mostly appressed to surface __________ **Placidium squamulosum**
      10. Thallus green, medulla K+ yellow or red, squamules upright __________ **Cladonia symphycarpia**

Appendix C: Species lists for keys and activities

Thallus Key

Acarospora contigua
Caloplaca flavovirescens
Cladonia peziziformis
Lichinella nigritella
Parmotrema reticulatum
Psora icterica
Punctellia bolliana
Ramalina sinensis
Teloschistes exilis
Xanthoparmelia lineola

Chemistry Key

Acarospora contigua
Buellia spuria
Caloplaca flavovirescens
Cladonia subtenuis
Cladonia symphycarpia
Physcia aipolia
Physcia biziana
Placidium squamulosum
Pleopsidium flavum
Punctelia bolliana
Usnea strigosa

Species with specific thallus features

Apothecia: *Physcia aipolia*
Cilia: *Teloschistes chrysophthalmus*
Isidia: *Placynthium nigrum, Myelochroa obsessa*
Perithecia: *Acarospora contigua*
Pigments for dyes: *Usnea* sp., *Parmelia* sp., *Ochrolechia* sp.
Podetia: *Cladonia* sp. except *C. apodocarpa*
Pruina: *Pyxine* sp.
Pseudocyphella: *Punctelia* sp.
Pycnidia: *Xanthomendoza* sp.
Soralia: *Parmotrema* sp. except *P. perforatum*

Species that give positive chemical test results

K+ *Physcia* sp., *Teloschistes* sp., *Xanthomendoza* sp., *Xanthoria* sp.
C+ *Diploschistes* sp., *Parmotrema tinctorum, Punctelia rudecta*
PD+ *Buellia dispersa, Flavoparmelia* sp.
UV+ *Acarospora contigua, A. socialis, A. novomexicana, Pleopsidiuum flavum*

Appendix D: Quick identification guide for common lichen genera of the region

Crustose thallus

Thallus color	Ascomatum type	Ascomatum color	Cells/spore	Spores/ascus	Special characters	Lower cortex	Substrate	Families & Genera
white or gray or yellow or brown	lecanorine, immersed	gray or yellow	1-celled	16–100+	UV+ bright orange	none	rock	**Acarosporaceae** *Acarospora* *Pleopsidium*
white-gray or green	perithecium-like or lecanorine, disks 1–3 per wart	pink-brown-black	1-celled or muriform in *Diploschistes*	1–8	warty thallus, some pruinose	none	clay, moss, or bark	**Graphidaceae** *Diploschistes* **Ochrolechiaceae** *Ochrolechia* **Pertusariaceae** *Pertusaria*
yellow	lecanorine disks	yellow to dark orange	2–4-celled	4–8	K+ purple	none	bark	**Teloschistaceae** *Caloplaca*
yellow	lecanorine disks	yellow to brown	1–2-celled	8–32	UV+ dark orange	none	bark	**Candelariaceae** *Candelariella*
green to gray	lecanorine or lecideine or biatorine	yellow-green or brown-black	1-celled	8		none	cone bracts, rock, or bark	**Lecanoraceae** *Lecanora* *Lecidella*
gray-white or brown or yellow	lecanorine disks some immersed some lecideine	brown-black, some pruinose	2–4-celled or muriform in *Diplotomma*	4–32		none	bark, wood, rock	**Physciaceae** *Buellia* *Dimelaena* *Rinodina*
white or gray-brown-black	perithecia	dark/black	1-celled or muriform in *Staurothele*	2–8	all chemistry tests negative	none	granite, limestone, gypsum, or shale	**Verrucariaceae** *Staurothele* *Verrucaria*

Endolithic thallus

Thallus color	Ascomata type	Ascomata color	Spore cells	Spores/ascus	Special characters	Lower cortex	Substrate	Families & Genera
white-brown-black	perithecia	brown-black	1–4-celled	8	all chemistry tests negative	none	rock	**Verrucariaceae** *Thelidium* *Verrucaria*

Fruticose thallus

Thallus color	Ascomata type	Ascomata color	Spore cells	Spores/ascus	Special characters	Lower cortex	Substrate	Families & Genera
black	wart-like thallinocarps	black	1-celled	8–32	ruffle-like lobes	black	rock	**Lichinaceae** *Lichinella*
yellow or green	lecanorine disks	yellow or brown	1-celled	8	slender branches	yellow or green	bark	**Parmeliaceae** *Evernia, Usnea*
yellow-orange	lecanorine disks	yellow-orange	polarilocular	8	cilia on terete branches or on apothecia of flat branches	yellow-orange	bark	**Teloschistaceae** *Teloschistes*
green or cream-brown	lecanorine disks	yellow-tan	2-celled	8	flat branches	green or cream-brown	bark	**Ramalinaceae** *Ramalina*

Squamulose thallus

Thallus color	Ascomata type	Ascomata color	Spore cells	Spores/ascus	Special characters	Lower cortex	Substrate	Families & Genera
dark gray to tan-brown	lecanorine, immersed	red-brown	1-celled	16–100	thallus concave or convex	rhizines or umbilicate	soil or rock	**Peltulaceae** *Peltula*
yellow-green or black or pink-tan-brown	biatorine	red-brown-black	1-celled	8		white-brown hyphae	soil or rock	**Psoraceae** *Psora*
green	biatorine caps or flat on podetia	red or light tan-brown	1-celled	8	squamules, or podetia, or both	various	soil or moss	**Cladoniaceae** *Cladonia*
white-gray or green or brown to dark brown	perithecia may be embedded	dark green-black	1-celled or muriform in *Endocarpon*	2 or 8	all chemistry tests negative	rhizines or pale hyphae or hairy	soil or rock	**Verrucariaceae** *Endocarpon* *Placidium*

Leprose thallus

Thallus color	Ascomata type	Ascomata color	Spore cells	Spores/ascus	Special characters	Lower cortex	Substrate	Families & Genera
green	sterile				no cortex, only soredia	none	bark, rock, or wood	**Stereocaulaceae** *Lepraria*

Foliose thallus

Thallus color	Ascomata type	Ascomata color	Spore cells	Spores/ascus	Special characters	Lower cortex	Substrate	Families & Genera
gray to green with pseudocyphella	lecanorine cups	tan-brown	1-celled	8	punctate with pseudocyphellae	light to black, rhizines	various	**Parmeliaceae** *Punctelia*
gray or yellow-green	lecanorine disks	yellow-tan-brown	1-celled	8		black-tan, variable rhizines, some white-brown-black rhizines	various	**Parmeliaceae** *Canoparmelia* *Flavoparmelia* *Hypotrachyna* *Myelochroa* *Parmotrema*
gray-green	lecanorine disks	brown-black	2-celled	8		light-tan-dark-black, variable rhizines	various	**Physciaceae** *Hyperphyscia* *Phaeophyscia* *Physcia* *Physciella*
gray-pruinose	lecanorine, but some with lecidine-like disks	black	4-celled	8	pruinose thallus	black, variable rhizines	various	**Physciaceae** *Pyxine*
yellow	lecanorine disks	orange	polarilocular	8		white, variable rhizines	various	**Teloschistaceae** *Xanthomendoza* *Xanthoria*
yellow	lecanorine disks	yellow	1-celled	20–50		white-tan, rhizines	bark	**Candelariaceae** *Candelaria*
gray-brown-black	lecanorine disks	red-brown	2–4-celled or muriform	4–8		gray-brown-black, hairy	soil or hardwood	**Collemataceae** *Collema* *Leptogium*
green/brown-black	embedded perithecia	black	1-celled	8		black-tan	rock	**Verrucariaceae** *Dermatocarpon*

Appendix E: Quick location guide for common lichen genera of the region

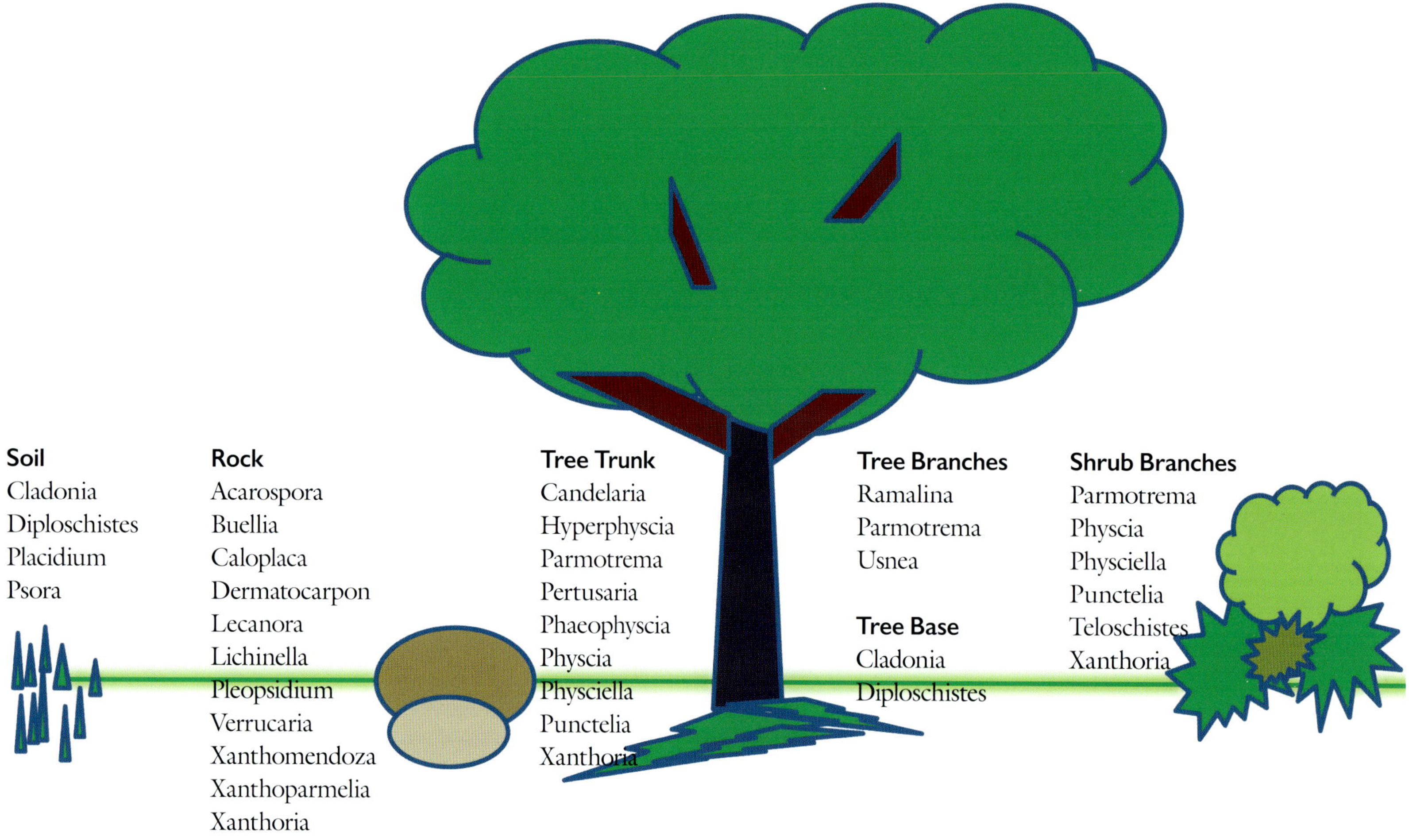

Appendix F: Websites of resources for lichen study

American Bryological and Lichenological Society
http://www.abls.org

Checklist of North American Lichens
http://www.ndsu.edu/pubweb/~esslinge/chcklst/chcklst7.htm

Consortium of North American Lichen Herbaria: Symbiotica
http://lichenportal.org/portal/collections/index.php

Federal Communications Commission, latitude/longitude conversions
https://www.fcc.gov/media/radio/dms-decimal

Kansas University, data and images of lichen collections
http://botany.biodiversity.ku.edu/botany/research/lichen-flora

Mycological Society of America
http://www.msafungi.org

New York Botanical Garden International Plant Science Center, C.V. Starr Virtual Herbarium
http://sciweb.nybg.org/science2/hcol/lena/index./asp.html

Opuscula phylolichenum, recent electronic publications in lichenology
http://sweetgum.nybg.org/science/vh/

Texas Tech University, images of lichens collected by Donna Maske
http://www.depts.ttu.edu/junction/LRFS/photos/lichens/

U.S. Department of Agriculture, descriptions and photos of lichen species
http://plants.usda.gov/java/

Bibliography

Brodo, I.M., S.D. Sharnoff, & S. Sharnoff. 2001. *Lichens of North America.* Yale University Press, New Haven, CT, U.S.A.

Brodo, I.M. 2016. *Keys to lichens of North America*: Revised and expanded. Yale University Press, New Haven, CT, U.S.A.

Consortium of North American Lichen Herbaria: Biodiversity occurrence data published by: <ASU, BRY, CMN, KANU, MIN, MSC, NY, OMA, SBBG, UCR, WTU> (Accessed through Consortium of North American Lichen Herbaria (CNALH) Data Portal, http://lichenportal.org/portal/index.php, 2015/08/01-30) http://lichenportal.org/portal/collections/index.php

Esslinger, T.L. 2015. A cumulative checklist for the lichen-forming, lichenicolous and allied fungi of the continental United States and Canada. North Dakota State University. (First Posted 1 December 1997, Most Recent Version (#20) 19 April 2015), Fargo, ND, U.S.A. http://www.ndsu.edu/pubweb/~esslinge/chcklst/chcklst7.htm

Hedrick, J. 1930. *Lichens from the state of Oklahoma*. Pap. Michigan Acad. Sci. 12:101–110.

McMullen, T. & F. Anderson. 2015. *Common lichens of northeastern North America*. Mem. New York Bot. Gard. Vol 112. The New York Botanical Garden Press, Bronx, NY, U.S.A.

Nash, T.H. III, ed. 1996. *Lichen biology*. Cambridge University Press, New York, NY, U.S.A.

Nash, T.H., C. Gries, & F. Bungartz, eds. 2004. *Lichen flora of the Greater Sonoran Desert Region*, Vol. II. Thomson-Shore, Inc., Dexter, MI, U.S.A.

Nash, T.H., C. Gries, & F. Bungartz, eds. 2007. *Lichen flora of the Greater Sonoran Desert Region*, Vol. III. Thomson-Shore, Inc., Dexter, MI, U.S.A.

Nash, T.H., B.D. Ryan, C. Gries, & F. Bungartz, eds. 2002. *Lichen flora of the Greater Sonoran Desert Region*, Vol. I. Thomson-Shore, Inc., Dexter, MI, U.S.A.

Purvis, W. 2000. *Lichens*. Smithsonian Institution Press, Washington, D.C., U.S.A.

Rankovik, B. 2014. *Lichen secondary metabolites, bioactive properties and pharmaceutical potential.* Springer International Publishing, New York, NY, U.S.A.

Showman, R. 2015. *Lichens of Ohio*. Ohio Division of Wildlife, Wildlife Diversity & Endangered Species Program Pub. 5473 (0115). Ohio Department of Natural Resources.

St. Clair, L.L. 1999. *A color guidebook to common Rocky Mountain lichens.* M.L. Bean Life Science Museum of Brigham Young University, Provo, UT, U.S.A.

USDA, NRCS. 2016. The PLANTS Database. http://plants.usda.gov (28 March 2016). National Plant Data Team, Greensboro, NC, U.S.A.

Webster, J. & R.W.S. Weber. 2007. *Introduction to fungi.* Cambridge University Press, New York, NY, U.S.A.

Will-Wolf, S. 2010. *Analyzing lichen indicator data in the forest inventory and analysis program.* United States Department of Agriculture Forest Service, Pacific Northwest Research Station, General Technical Report PNW-GTR-818. September.

Index of Lichen Species Names and Photographs

ACTUAL SCALE

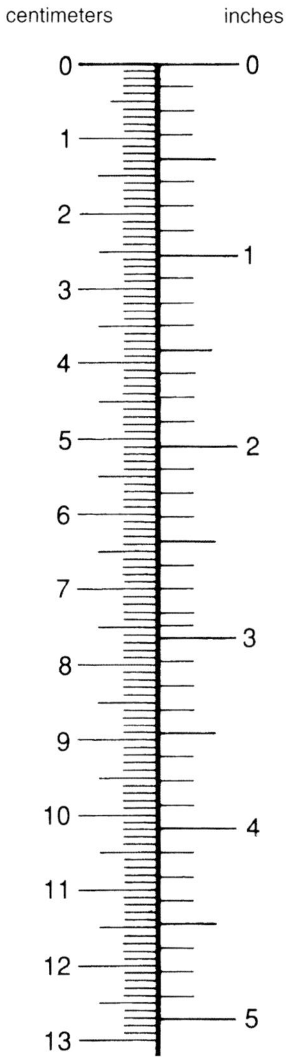